WILDF1
ZION NATIONAL PARK

Common Wildflowers, Ferns and Cacti of
Zion and Bryce Canyon National Parks and
Cedar Breaks National Monument

SEGO-LILY (*Calochortus nuttallii*)

STEVE CHADDE

Botanical (scientific names) in this guide follow those of ITIS, the Integrated Taxonomic Information System (*www.itis.gov*). Photographs were obtained from the author's collection, public domain sources, and a number of photographers who have made their images available under commercial-use licences on Flickr (*www.flickr.com*).

Enjoy your visit to our nation's national parks and monuments, but remember, all plant life is protected in national parks, and please refrain from picking any of the plants you see in these areas.

WILDFLOWERS OF ZION NATIONAL PARK
Common Wildflowers, Ferns and Cacti of Zion and Bryce Canyon National Parks and Cedar Breaks National Monument

Steve Chadde

 A Pathfinder Field Guide, published by Orchard Innovations

ISBN 978-1-951682-52-1

CONTENTS

FRINGED GENTIAN (*Gentianopsis thermalis*)

Southern Maidenhair Fern (*Adiantum capillus-veneris*), growing in a seepage area of a cliff face, Zion Canyon.

INTRODUCTION

Wildflowers of Zion National Park is a non-technical introduction to over 100 of the more common wildflowers found in Zion, as well as in nearby Bryce Canyon National Park and Cedar Breaks National Monument. In addition, common ferns (pages 7–15) are described as are six of the region's most common cacti (pages 115–121). Following the ferns, wildflowers are arranged by flower color, then by plant family, so that related species are grouped together. The Index (page 123) includes common names, scientific names, and synonyms (other formerly used scientific names). A special index to each plant family is provided on page 122.

In the region of Zion and Bryce Canyon National Parks and Cedar Breaks National Monument, four general life zones are found extending from low elevations of near 3,600 feet to extremes of over 11,300 feet. The area from the lowest point to 4,000 feet is the **Lower Sonoran Zone**; above that to 7,000 feet is the **Upper Sonoran Zone**; from 7,000 to 8,500 feet is the **Transition Zone**; above that to 10,000 feet is the **Canadian Zone**; and from 10,000 feet to the highest point on Brian Head Peak is the **Hudsonian Zone**. Native plants typical of deserts, mesas, and mountains grow within these extremes of elevation, and provide a home for a wide variety of species. For example, more than 1,000 plant species have been found within Zion National Park, and of the three areas featured, Cedar Breaks National Monument contains the greatest diversity, resulting in marvelous displays of wildflowers, depending on the amount of rainfall from year to year.

Vegetation and Climate

The vegetation of Zion National Park is an interesting mix of desert and mountain flora, due to the contrasting landscapes within the Park, which is located on the boundary between the high cool plateaus of Utah and the hot desert of the southwest. Here, at the southern edge of the Markagunt Plateau, the Virgin River (a tributary of the Colorado) has cut deep canyons that open to the deserts to the south; resulting in three well-defined environments, each with a distinctive climate and vegetation: cool plateaus at 6,500 to 7,500 feet elevation in the north; a small strip of hot desert below 4,000 feet elevation in the south; and a network of canyons between, which have a climate intermediate between the plateaus and desert.

The plateaus, including numerous flat-topped peaks and mesas, are covered with chaparral and forests in which ponderosa pine is the dominant tree, with Douglas-fir and quaking aspen in moister places. This is the **Transition Zone**, or **Ponderosa Pine Belt**; in other words, the climate and vegetation is similar to that found at low altitudes in the northern states where there is a transition between a hot southern climate and the colder conditions of Canada.

In the canyons, between 4,000 and 7,000 feet elevation, the climate is somewhat warmer, usually too warm for ponderosa pine, but favorable for such trees as Utah juniper and the pinyon pines, which form a sparse 'pygmy forest' on all the dry slopes and flats between these altitudes. Along streambanks and other moist places in the canyons, Douglas-fir and quaking aspens typical of the plateaus are usually replaced by cottonwoods, ashes, and boxelders. This belt of climate and vegetation is termed the **Upper Sonoran Zone.**

Along the southern border of the Park, where the canyons open out upon relatively flat, semi-desert country, the pygmy forest gives way to desert chaparral, in which creosotebush and blackbrush are the dominant shrubs. This is known as the **Lower Sonoran Zone**, and although only a few square miles of park territory are within this zone, it is important botanically because of the large number of species unique to it, and because of the interesting ways in which many of these species survive the extreme heat and drought.

These three climatic zones can be easily recognized by any Park visitor. Desert shrubs indicate **Lower Sonoran Zone**, pygmy forest of pinyon and juniper indicates **Upper Sonoran Zone**, and ponderosa pine indicates the **Transition Zone**. The vegetation is so naturally divided into these three groups that in this guide the distribution of many plants is given simply by zones. In addition, at Cedar Breaks, elevations are higher, and above the Transition Zone at 8,500 feet to 10,000 feet is the **Canadian Zone**; and from 10,000 feet to the highest point on Brian Head Peak (11,312 feet) is the **Hudsonian Zone.**

However, one should not think that as soon as they pass above an elevation of 7,000 feet on a trail or road, they will abruptly pass from scrub junipers to stately ponderosa pines. The change is always gradual, and may occur 1,000 feet or more above or below the 7,000-foot average. In a shaded canyon the climate will be much cooler at 5,500 feet than at 6,500 feet on a sunny south slope. For example, in going up the West Rim Trail, one finds ponderosa pine and quaking aspen growing in a shady spot above Scout Lookout at 5,600 feet, while at 7,200 feet, on the south slope of Horse Pasture Plateau, they will pass through open woodlands of pinyon and juniper. The complex system of canyons in Zion causes many such reversals of the normal zonal distribution of plants. In this region, the life zone idea should be considered as the general pattern and where intricate variations have been produced by differences in sunlight, moisture, and soil.

The most noticeable of these variations is to be seen in the bottom of nearly every canyon, where additional moisture and shade favor the growth of Transition or even Canadian Zone species, regardless of the actual altitude. In canyons where water seeps continually from the cliffs, there are great "hanging gardens" of ferns (the most common of which are described in this guide) and other plants that would normally be found at a much higher altitude. The Narrows is the classic example of this, but dozens of other canyons have the same type of vegetation.

Aside from zonal distribution, the climate produces another, entirely different phenomena — a yearly cycle of blooming periods. The additional moisture and moderate temperature of spring produces the most varied flower display of the year during March, April, and May. In early summer, as the result of a hot drought period typical of June, there are only a few species in bloom, and these are mostly night-bloomers. From August to October, the late summer thunderstorms cause a second flower display nearly equalling that of the spring months.

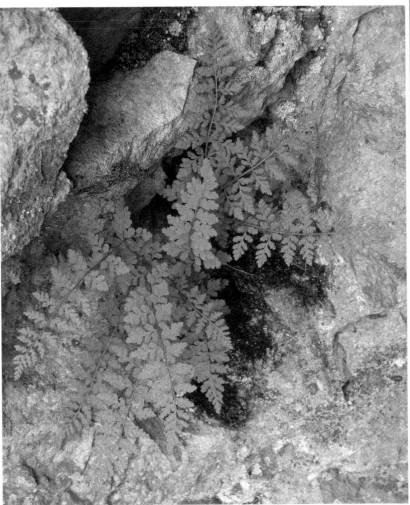

BRITTLE FERN
Cystopteris fragilis   CYSTOPTERIDACEAE | BLADDER FERN FAMILY

Brittle Fern is common in all the cool canyons, growing on moist slopes and ledges, and in crevices.

BRACKEN FERN

Pteridium aquilinum   DENNSTAEDTIACEAE | BRACKEN FERN FAMILY

The **Male Fern** and the **Bracken Fern** are the only large, coarse ferns in the park, but Bracken Fern is easily distinguished by its typical habitat of fairly open slopes or flats having deep soil, rather than in canyons. Also, it does not grow in clumps, but as single, erect fronds, on a stout "stem" (stipe). It has a very coarse appearance, and grows from 1 to 3 feet high.

WESTERN POLYPODY

Polypodium hesperium POLYPODIACEAE | POLYPODY FERN FAMILY

In the coolest canyons **Western Polypody**, the plainest appearing of our small ferns, is occasionally found growing in the same moist places as the **Fragile Fern**. Visitors from the Pacific Northwest will recognize it as a relative of the **Licorice Fern** (*Polypodium glycyrrhiza*). It is most apt to be seen in Hidden Canyon and some of the small canyons around Mount Majestic on the West Rim Trail, but is not common.

MALE FERN

Dryopteris filix-mas　　　　DRYOPTERIDACEAE | WOOD FERN FAMILY

Male Fern is by far the largest fern that will be found in the cool, moist canyons; its rank, upright fronds reach a length of from 1 to 4 feet. It has heavy, tapering fronds ascending from a compact crown, and is not likely to be confused with any other fern in the Park. It is common in nearly every cool canyon at elevations above 5,500.

MAIDENHAIR FERN

Adiantum aleuticum PTERIDACEAE | MAIDENHAIR FERN FAMILY

SYNONYM *Adiantum pedatum* subsp. *aleuticum*

Maidenhair Fern is perhaps the most beautiful and best known of our ferns. Found in nearly all the cool canyons of Zion, but much less common than the Southern Maidenhair Fern (next page).

SOUTHERN MAIDENHAIR FERN

Adiantum capillus-veneris PTERIDACEAE | MAIDENHAIR FERN FAMILY

Southern Maidenhair Fern is very common in all cool, moist places in the park, especially in the hanging gardens at Weeping Rock and The Narrows, where it grows profusely on wet cliffs.

SCALY LIPFERN

Cheilanthes covillei   PTERIDACEAE | MAIDENHAIR FERN FAMILY

SYNONYM *Myriopteris covillei*

Scaly Lipfern is odd little fern that grows on dry rocks and in crevices, but is nowhere common. In dry weather its fronds curl up into small round balls with the scaly side exposed. Following rain, the fronds uncurl once more.

SMOOTH CLIFFBRAKE

Pellaea glabella　　　　　PTERIDACEAE | MAIDENHAIR FERN FAMILY

All of the **Cliffbrakes** are small ferns of dry, rock habitats, and all are rather rare here. This species is known from the foot of the West Rim Trail and Temple of Sinawawa, Zion Canyon. The related **Spiny Cliffbrake**, *Pellaea truncata*, has been found in dry, rocky places near the base of Angel's Landing, and in dry talus near Emerald Pool.

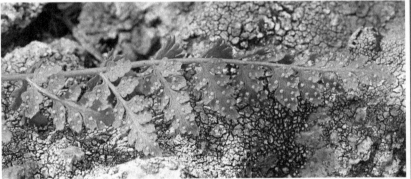

CLIFF FERN

Woodsia oregana   WOODSIACEAE | CLIFF FERN FAMILY

Cliff Fern is a small, delicate fern, having a resemblance to Brittle Fern. It is known from rock crevices in Hidden and Echo Canyons, and may occur in other similar habitats.

Zion National Park is home to more than 1,000 plant species. Shown here are **Indian-Paintbrush** (*Castilleja* spp.), **Desert Sage** (*Salvia dorrii*), **Desert Prince's-Plume** (*Stanleya pinnata*, far right), as well as several species of grass.

FOURWING SALTBUSH

Atriplex canescens   AMARANTHACEAE | AMARANTH FAMILY

Fourwing Saltbush is a silvery-green, profusely branched shrub, growing two to five feet high, with conspicuous clusters (in late summer) of four-winged seeds about the same color as the leaves. It is very common in the Sonoran Zones, and abundant in the alkaline flats of the Great Basin of Utah. In the lower portion of Zion Canyon it is abundant and often mistaken for **Sagebrush** (*Artemisia*), which it resembles to some extent. This plant is of value as forage for livestock, and deer feed upon it to a limited extent.

GREEN EPHEDRA

Ephedra viridis EPHEDRACEAE | MORMON-TEA FAMILY

Green Ephedra is not a very showy plant with its pale-green stems, very small leaves and inconspicuous flowers. It is probably of greatest interest because of the use made of the plant by early pioneers in brewing a tea, which served as a tonic for various ailments. It was commonly called **Brigham Tea**, **Squaw Tea** or, more generally, **Mormon-Tea**. The plant is a relative of the Pines and Firs and is able to withstand drought. It is found fairly abundantly in the Sagebrush and Saltbush areas of the Upper Sonoran Zone. The drug ephedrine is obtained from some of the species of *Ephedra* found in China.

WOODLAND PINEDROPS

Pterospora andromedea   ERICACEAE | HEATH FAMILY

In the rich soil of the Ponderosa Pine forest of the high plateaus you may occasionally see this peculiar plant called **Woodland Pinedrops**. It has a single reddish-brown stalk (no green parts at all) about a foot or two high, apparently without leaves (they are reduced to scales), and numerous round or bell-shaped blossoms or seed pods hanging from short stems. The stalk is generally sticky with a material much like the pitch of the pine trees. It is a parasite that lives on the roots of pine trees.

GREEN GENTIAN

Frasera speciosa

GENTIANACEAE | GENTIAN FAMILY

Green Gentian is found in the open meadows of the Cedar Breaks highlands and on the high plateaus of Zion and Bryce Canyon. The tall stalks, with their intermixture of pale-green leaves and similarly colored flowers, present a rather conspicuous sight as they grow to heights of 5 feet. The flowers have four sepals and four petals and nectar glands that attract many insects. The petals are flecked with brown and purple.

ELDERBERRY

Sambucus racemosa   ADOXACEAE | MUSKROOT FAMILY

This red-fruited **Elderberry** is found most abundantly at Cedar Breaks. It is also common along the highways through forested areas of the region. The clusters of small white flowers that come on usually in June or July give way to bright-red berries in August and present a most attractive display. Most Elderberries are edible and are eaten by birds and rodents. Some people gather the berries for wines and jellies. This species, however, is considered poisonous, and cases of poisoning have been reported from eating the berries, flowers, roots and bark. The stalks of some Elderberries are pithy and fairly easily hollowed-out. Native Americans used the stalks for making a simple flute.

FINELEAF YUCCA

Yucca angustissima   AGAVACEAE | AGAVE FAMILY

The name "Our Lord's Candlestick" was given to this tall, conspicuous desert plant by the early Spanish padres, who were the first white men to visit this region of southern Utah. During May and June the waxy-white flowers bloom on tall stalks and soon mature into rather large seed pods. Native Americans made very good use of all parts of the plant: its fiber was used for making sandals and clothing, the seeds provided food, and the roots were used for making soap. The Navajo called it 'Yaybi-tsa-si', which literally means "Yucca of the Gods."

INDIAN-POTATO

Orogenia linearifolia APIACEAE | CARROT FAMILY

As the snow melts on the plateaus, at elevations above 7,000 feet, you may find **Indian-Potato**, one of the very first flowers of spring. This member of the Carrot Family often carpets alpine meadows with a mass of tiny white flowers never more than an inch or two high. The root bulb was eaten by Native Americans and gives rise to the common name Indian-Potato. Another common name is **Salt and Pepper Plant** because of the speckled appearance of the flowers. The blooming period of this plant is very brief, and soon after the flowers have faded, the leaves disappear, and the plant lies dormant during most of the year.

PORTER LIGUSTICUM

Ligusticum porteri APIACEAE | CARROT FAMILY

Locally called **Wild Parsley**, this fairly tall plant, with its fern-like leaves, is very common at Cedar Breaks National Monument. It grows at very high elevations. Other plants in this family, useful as food, are the carrot, parsnip, dill and anise. However, one member of the family known from the region, **Water Hemlock** (*Cicuta maculata*), is very poisonous, and often fatal if eaten.

WESTERN YARROW

Achillea millefolium ASTERACEAE | SUNFLOWER FAMILY

SYNONYM *Achillea lanulosa*

Western Yarrow is very widespread and can be found inearly everywhere in North America. Here, it is more common at elevations above 5,000 feet. It grows 12 to 20 inches high, and the fernlike leaves, which have a pungent odor when crushed, and the umbrella-shaped clusters of flowers, are characteristics of this plant that help to identify it. Since ancient times, the plant has been highly regarded for its healing properties. Legend ascribes the discovery of this virtue to Achilles, in whose honor the plant is named.

BITTERCRESS

Cardamine hirsuta   BRASSICACEAE | MUSTARD FAMILY

You may find **Bittercress** blooming during April and May in the Sonoran Zones of Zion National Park. Its habitat is generally dry, sandy hillsides rather than in the deep canyons. The wide-spreading, circular clumps of leaves are topped with an attractive display of pure white flowers. The petals of four are arranged like a cross. Being a perennial, the clumps expand from year to year, and often reach a spread of four to five feet across. The plants are useful in building soil and in preventing erosion.

WHITE-MARGIN EUPHORBIA

Euphorbia albomarginata EUPHORBIACEAE | SPURGE FAMILY

This low-growing plant with abundant small, round or oval leaves and tiny white flowers is commonly called **Carpetweed** because of the manner in which it spreads over the ground. Found mostly in dry sandy soils, this plant serves as an excellent soil -inder and is helpful in preventing erosion. The milky juice of this species is considered toxic and may cause dermatitis in susceptible people. It is also known as **Rattlesnake Feed** and popularly supposed to be efficacious in treating snake-bites.

SEGO-LILY, MARIPOSA LILY

Calochortus nuttallii

LILIACEAE | LILY FAMILY

The **Sego-Lily** was chosen as the State Flower of Utah probably because of the important use early pioneers made of theplant's root bulbs in supplementing their meager diets during the early settlement period. The name "Sego" (pronounced *see-go*) is of Shoshonean origin, and this flower was sacred in Native American legend long before the arrival of Utah pioneers in 1847. This plant is found in rather dry, rocky soils, and puts on a very showy display during May and June. It is commonly known as the **Mariposa Lily** in other sections of the country, but in Utah it is usually called **Sego-lily**. In the Zion region there are five *Calochortus* species with flowers varying from white, pink, and yellow.

MOUNTAIN DEATHCAMAS

Anticlea elegans　　　　MELANTHIACEAE | FALSE HELLEBORE FAMILY

SYNONYM *Zigadenus elegans*

Mountain Deathcamas is an onion-like plant, with a long, loose cluster of small, creamy-white flowers. Its root is shaped much like that of the onion, but is odorless. *The plant is poisonous to people and animals, and should never be eaten.* **Deathcamas** is found mostly in meadows or wet places on the plateaus where it presents a serious danger to grazing cattle and sheep. At Cedar Breaks it blooms during July and August, and is fairly common in alpine meadows.

SPRINGBEAUTY

Claytonia lanceolata MONTIACEAE | CANDY-FLOWER FAMILY

Springbeauty is abundant in Cedar Breaks during May and early June, and also in the shady canyons of Zion in the Transition Zone. Plant have two narrow leaves near the base, each about 2 inches long, above which are four to five practically leafless branches with a single flower at the top of each. The plant is rarely over 6 inches high, and more commonly it is flattened on the ground. Flowers vary in color from white to pink or sometimes the white blossoms have pink veins or stripes. Springbeauty is usually one of the early blooming flowers of the high plateaus, along with the **Indian-Potato** and **Buttercup.**

Lewisia rediviva (above), *Lewisia pygmaea* (left).

BITTERROOT

Lewisia rediviva   MONTIACEAE | CANDY-FLOWER FAMILY

This beautiful, dwarfed plant, never more than an inch or two high, is found during May, mostly on the lava fields of the Transition Zone in Zion National Park. Its flower of white petals with pink veins is about 2 inches in diameter. **Bitterroot** was discovered in 1805 by the Lewis and Clark expedition while passing through western Montana. It was later named *Lewisia rediviva* by the botanist Pursh, and in 1895 was named the official state flower of Montana. The plant is of economic importance to northwestern Native Americans, who discovered that the bitter, parsnip-shaped roots have a nutritious heart of starch, which cooking reduces to a pasty mass, palatable, at least, to Native American taste. Also present in the region is the similar **Pygmy Bitterroot**, *Lewisia pygmaea,* and hybrids between the two species are reported.

WHITE EVENING-PRIMROSE

Oenothera cespitosa   ONAGRACEAE | EVENING-PRIMROSE FAMILY

White Evening-Primrose, with its fragrant, large white flowers, blooms early in the evening and lasts only a few hours the next day, before turning pink and wilting. The stemless flowers grow close to the ground, and have long, slender, and hairy calyx tubes that resemble stems. This plant is found mostly in very rocky and dry habitats, which are often devoid of nearly all other plants.

PRICKLYPOPPY

Argemone munita   PAPAVERACEAE | POPPY FAMILY

You can find **Pricklypoppy**, with its large, white flowers, most frequently along road-cuts or in abandoned fields where it has taken over as a weed. Its showy display is most abundant during midsummer or in July and August. The large, white flower, with its conspicuous yellow center, is sometimes called "The Cowboy's Fried Egg." The prickly leaves and stems account for another common name, **Thistle-Poppy**. The plants are drought resistant and unpalatable to livestock. The seeds contain a narcotic drug similar to opium.

HOARY PHLOX, CARPET PHLOX

Phlox canescens   POLEMONIACEAE | PHLOX FAMILY

SYNONYM *Phlox hoodii* subsp. *canescens*

Several kinds of **Phlox** are common in Zion, and other species are found at Cedar Breaks and Bryce Canyon. The plants are generally small, but the flowers are fairly showy with their five white (or also pink or bluish) petals. They are found generally in dry, rocky soils, and bloom mostly in early summer or during May and June. Sometimes Phlox will cover extensive areas forming a bright carpet of pink or white; a display of Phlox may be found on the summit of Brian Head Peak near Cedar Breaks. These plants are very helpful in holding the soil and in preventing erosion.

WESTERN BANEBERRY

Actaea rubra   RANUNCULACEAE | BUTTERCUP FAMILY

Western Baneberry may be found in moist, shady forests. It grows to about one to two feet tall, with handsome leaves, but with rather inconspicuous heads of creamy-white flowers. Later, the conspicuous bright-red or waxy-white berries about the size of peas in a dense cluster make a very attractive display in late summer. The berries are somewhat poisonous, hence the name **Baneberry**.

SERVICEBERRY

Amelanchier alnifolia   ROSACEAE | ROSE FAMILY

One of the very early blooming shrubs or small trees in Zion Canyon, and often seen in March or April as conspicuous white masses on the canyon walls among the Pinyon and Juniper trees, is the **Serviceberry**. The white flowers occur in clusters and look much like a fruit tree in bloom. The berry is shaped like a very small apple, insipid to the taste, but nevertheless used extensively by Native Americans and early settlers for food. The plant has a variety of common names, including the following: **Juneberry**, **Shadbush**, **Saskatoon**, **Sarviceberry** and **Pigeonberry**.

CHOKECHERRY

Prunus virginiana

ROSACEAE | ROSE FAMILY

Chokecherry bushes or small trees are fairly common at elevations around 6,000 feet. Early in May, and sometimes even earlier, the white, fragrant blossoms present a pleasing sight. In July or August the ripened cherries supply food for humans, and many birds and small animals. The first autumn colors are often the Chokecherry leaves as they turn scarlet, brown, and red. These shrubs are browsed heavily by cattle and somewhat by deer, especially if other forage is scarce.

STANSBURY CLIFFROSE

Purshia stansburiana   ROSACEAE | ROSE FAMILY

SYNONYM *Cowania stansburiana*

In Zion this shrub is usually found at elevations above 4,000 feet, and generally grows to 6 to 8 feet high. During May and June its creamy-white flowers are suggestive of the **Wild Rose**. The habitat of the **Cliffrose** is the warm, dry slopes among the Pinyon and Juniper. Its twigs and evergreen leaves are browsed extensively by deer and other animals. The fragrance of this plant in bloom is remarkable, and reminds one of orange blossoms. The fruits are achenes, with pale, silky-haired tails 2 or more inches in length, which at times are very showy.

SACRED DATURA

Datura wrightii

SOLANACEAE | POTATO FAMILY

SYNONYM *Datura meteloides*

Sacred Datura is a conspicuous plant with very large, white (sometimes lavender-tinted) flowers that open at night and wilt in the bright morning sunlight. A single plant may have as many as 100 blooms at one time. *Datura* is one of the few plants that blooms during the hot summer in Zion Canyon. Many different names are locally applied to it, including: **Zion Lily, Moon Lily, Jimson Weed** and **Thornapple**. The plant is poisonous to eat, especially the seeds, and was used by several Native American tribes to induce stupor and dreams as a part of a widespread religious cult. It contains a deadly narcotic principle akin to atropine.

CANADIAN WHITE VIOLET

Viola canadensis   VIOLACEAE | VIOLET FAMILY

The most common **Violets** in the region are *Viola adunca, Viola canadensis, Viola nephrophylla,* and *Viola palustris.* Flower color ranges from white to yellow or purple. They are found blooming in May or June on the high plateaus or sometimes later in the wet meadows of Cedar Breaks or in shady canyons. Violets are closely related to the cultivated Pansies. The flowers are irregular, as one petal has a saclike spur which contains nectar. They have five sepals, five petals and five stamens. This white-petaled species and the purple-flowered types are most common in very moist and cool areas of the shady canyons. Yellow-flowered Violets are found most commonly in shaded forest.

BLUEDICKS

Dichelostemma capitatum   ASPARAGACEAE | ASPARAGUS FAMILY

SYNONYM *Dipterostemon capitatus*

This bright blue-purple flower, on its long, slender stalk, has a number of common names, such as **Wild Hyacinth**, **Grass Nuts** and **Spanish Lily**. The bulb of this plant has a nutty flavor. It was gathered by Native Americans and early pioneers for food in some sections of the country. It is found sparingly in the region's habitats having rich soil.

AMERICAN-ASTER

Symphyotrichum spp.

ASTERACEAE | SUNFLOWER FAMILY

SYNONYM *Aster* spp.

The **American-Asters** and **Fleabanes** are sometimes confused, but they can generally be recognized by the difference in the number of ray flowers – American-Asters have only about half as many ray flowers as do the Fleabanes. Species of American-Aster form an important part of the late summer floral display at Cedar Breaks and Bryce Canyon. They come on after the **Lupine, Columbine, Indian Paintbrush** and other early bloomers have faded. Our Asters are now placed in genus *Symphyotrichum* (mostly endemic to North America), as members of the genus *Aster* now refers to species nearly all of which are restricted to Eurasia. Identification of species can be challenging; illustrated above is *Symphyotrichum ascendens*, one of the region's more common species.

MEADOW SALSIFY

Tragopogon porrifolius   ASTERACEAE | SUNFLOWER FAMILY

This interesting plant, also known as **Oyster Plant**, has been naturalized from Europe and is now quite common in the West. It has a smooth, stout, hollow stem about 2 feet tall; rather dark-green, smooth leaves clasping at the base; and handsome flowers from 2 to 4 inches across. The flowers open early in the morning, close at midday, and remain closed until the next morning. As the flower matures into seeds in a conspicuous and very large, dandelion-like head, each seed is equipped with a perfect parachute of silky fibers; winds often sweep these flight-equipped seeds for many miles. **Meadow Salsify** is most commonly found along roadways or in other places where the native soil has been disturbed.

MOUNTAIN BLUEBELLS

Mertensia arizonica　　　　　BORAGINACEAE | BORAGE FAMILY

Mountain Bluebells is a fairly tall perennial growing to 2 to 3 feet high, and found in moist places at high elevations of the plateaus. It is common at and near Cedar Breaks along streams or in swampy meadows. The small clusters of drooping, delicate blue flowers come mostly in May and June, or even later at the higher elevations of Brian Head Peak near Cedar Breaks. Before maturing the flowers may be pinkish to white in color.

AMERICAN HAREBELL

Campanula rotundifolia   CAMPANULACEAE | BELLFLOWER FAMILY

In the drier habitats of the alpine regions around Cedar Breaks, and generally along the roadsides in large clumps, you may find the beautiful blue flowers of **American Harebell**; the deep-blue flowers, drooping on their hairlike stems, have such perfect shape and simple grace. The Harebell is very widespread, being found in Scotland, northern Europe and Asia, and over much of North America.

PRAIRIE SPIDERWORT

Tradescantia occidentalis COMMELINACEAE | SPIDERWORT FAMILY

In sandy areas at elevations above 4,000 feet you may find this pretty, three-petaled, deep-blue or purplish flower, on its slender stalk about a foot or more in height, blooming early in June. The flowers bloom at night, so are not easily found except early in the mornings. The plant is fairly abundant along the Narrows Trail, the East Rim Trail, and near the East Entrance Station of Zion National Park. Native Americans used the entire plant for food.

DESERTBEAUTY DALEA

Psorothamnus fremontii

FABACEAE | PEA FAMILY

SYNONYM *Dalea johnsoni*

Sometimes confused with **Desert Sage** or more commonly called **Purple Sage**, which it resembles to some extent, this small shrub with light-gray bark, small, gray-green leaves and terminal spikes of brilliant purple flowers is one of the most pleasing sights in early Summer. In Zion it is found mostly in the Coalpits Wash and Shunes Creek areas, and generally blooms in May. **Desertbeauty Dalea** is a close relative of **Smokebush** (*Psorothamnus spinosus*) of low-elevation desert.

FRINGED GENTIAN

Gentianopsis thermalis

GENTIANACEAE | GENTIAN FAMILY

SYNONYM *Gentiana thermalis*

One of the most beautiful of all mountain flowers, **Fringed Gentian** is commonly found in the moist meadows of Cedar Breaks at high-elevations (near or above 10,500 feet). The flower stalks are generally 6 to 10 inches tall, and each bears a handsome flower about two inches long, with four fringed petals. At times **Fringed Gentian** may carpet alpine meadows with a waving mass of deep-blue color. This species is the 'Park Flower' of Yellowstone National Park.

DESERT SAGE

Salvia dorrii   LAMIACEAE | MINT FAMILY

Although looking very much like a clump of Sagebrush, this small shrub is a member of the Mint Family, and not closely related to Big Sagebrush (*Artemisia tridentata*). The clusters of bright purple flowers, blooming in May or June, brilliantly adorn this sage-green plant. It is fairly common in the Sonoran Zones, and well-scattered throughout much of the Southwest. This is the plant referred to in the storied book by Zane Gray, *Riders of the Purple Sage*. One of the best places to look for this plant is along the Emerald Pool Trail in Zion National Park.

WILD FLAX, LEWIS FLAX

Linum lewisii   LINACEAE | FLAX FAMILY

In Zion during May and June, growing along the trails on the plateaus and in cool canyons, you will find the beautiful and delicate blue flowers of the **Wild Flax**. The flower is nearly an inch across and has five sepals and five petals borne at the top of a slender stem having narrow leaves. At Bryce Canyon this plant is more abundant than at Zion or Cedar Breaks. Wild Flax was named in honor of Captain Meriwether Lewis, who first discovered it near the Continental Divide, during the famed Lewis and Clark expedition of 1804-1806.

SILKY LUPINE

Lupinus sericeus

FABACEAE | PEA FAMILY

There are so many varieties of **Lupine** that it is most difficult to identify the numerous species. In this area they are found abundantly on the high plateaus, being especially plentiful at Cedar Breaks, where they fill whole meadows with a mass of blue color in mid-summer. Lupines range in color from pale pink to deep purple, to white, cream, or yellow, but most are blue-flowered. Like other plants of the Pea Family, Lupines add nitrogen to the soil and thereby improve the land on which they grow. The seeds of a few species contain alkaloids which are poisonous to livestock, especially sheep.

ELEPHANTHEAD PEDICULARIS

Pedicularis groenlandica   OROBANCHACEAE | BROOM-RAPE FAMILY

You will find this strange-looking plant in the wet meadows of the alpine areas of Cedar Breaks, and on the Plateaus of the Kolob Section and the Horse Pasture Plateau of Zion National Park. Its blooming season is July and August, and its flowers vary from purple to pink. The peculiarly modified petals of the corolla resemble the forehead, ears and waving trunk of an elephant, hence the common name **Elephanthead**.

DUSTY PENSTEMON

Penstemon comarrhenus   PLANTAGINACEAE | PLANTAIN FAMILY

Dusty Penstemon is generally common along roadsides at elevations above 4,500 feet. It is fairly common at Cedar Breaks and at Bryce Canyon. The flowers vary in color from deep blue to dark purple, and the stalks vary in height from 12 to 20 inches.

THICKLEAF PENSTEMON

Penstemon pachyphyllus   PLANTAGINACEAE | PLANTAIN FAMILY

The penstemons are sometimes called **Wild Snapdragons** because of the close resemblance to related cultivated species. They are also called **Beardtongue** because one of the five stamens is covered with numerous hairs. This blue-flowered species, **Thickleaf Penstemon**, is found mostly at higher elevations or on the plateaus, where they bloom during June and July.

MONKSHOOD

Aconitum columbianum   RANUNCULACEAE | BUTTERCUP FAMILY

Monkshood is found abundantly at Cedar Breaks in more open forested areas where there is partial shade and plenty of moisture. The purple Monkshood rank almost as high as their cousins the Columbines and Larkspurs in charm and beauty, with a quaintness and individuality all their own. The flower features a modified sepal shaped like a hood or helmet that covers the stamens. Extracts from the thick, turnip-shaped root are used medicinally in the treatment of certain heart diseases.

BLUE COLUMBINE

Aquilegia caerulea   RANUNCULACEAE | BUTTERCUP FAMILY

The beautiful flowers of **Blue Columbine** are well-known because of their wide distribution and common use as cultivated species in flower gardens. The flowers, which can vary from white, yellow, pink, red, or blue in related species, have conspicuously shaped petals with long, hollow spurs, which contain honey and thereby attract certain insects and especially humming-birds. **Columbines** are probably the most beautiful of the native flowers of Zion, Bryce Canyon and Cedar Breaks; their attractive displays can be seen during the summer months.

NUTTALL'S LARKSPUR

Delphinium nuttallianum   RANUNCULACEAE | BUTTERCUP FAMILY

Larkspurs are found abundantly during July and August in the alpine meadows of Cedar Breaks Monument. The leaves are very similar to those of the **Monkshood** (*Aconitum*), but the flowers differ in color and shape. The single spur of one of the sepals is the marked feature of the Larkspur. Color and size vary greatly for the different species; but **Nuttall's Larkspur** (common at Cedar Breaks) is 2 to 3 feet tall, and the flowers are a purplish blue.

ELK THISTLE

Cirsium foliosum   ASTERACEAE | SUNFLOWER FAMILY

Elk Thistle is widely scattered in the region but is not very abundant. It is a stout plant, 2 to 3 feet tall, with large, prickly leaves. Its freshly budding, rich pink flowers are very attractive during the early summer. Hummingbirds and numerous insects gather food from its colorful flower-head made up of many individual flowers. Other thistles, such as **Canada Thistle** (*Cirsium arvense*), are obnoxious weeds and detrimental to gardeners and agriculture.

FLEABANE

Erigeron speciosus   ASTERACEAE | SUNFLOWER FAMILY

There are numerous species of **Fleabane** in this region. Some particular kind may be found in flower at any time of the growing season, for certain species bloom very early and others continue late in autumn. Some species of Fleabane grow in dense masses and, in early spring, carpet the meadows and roadsides. The ray flowers of the Fleabanes are generally twice as numerous per head as are the ray flowers of the **American-Asters**. The plants are quite similar in other respects.

SNOWBERRY

Symphoricarpos oreophilus CAPRIFOLIACEAE | HONEYSUCKLE FAMILY

SYNONYM *Symphoricarpos utahensis*

This low, spreading shrub is recognized by its shreddy bark; small, oval, opposite leaves on very short petioles, and in late summer or fall by its white berries. The small, pinkish flowers are not at all conspicuous and are often overlooked. The plant is browsed by deer and other animals and is sometimes called **Buckbrush**. It is found mostly on the high plateaus of Zion and is fairly common at Cedar Breaks and Bryce Canyon. The fruit, although very showy, is insipid and not very tasty.

ROCKY MOUNTAIN BEEPLANT

Peritoma serrulata   CLEOMACEAE | SPIDER-FLOWER FAMILY

SYNONYM *Cleome serrulata*

Along the roadsides of southern Utah near Zion, Bryce Canyon and Cedar Breaks, you may find this attractive floral display of rich pink or purple presented by patches of **Rocky Mountain Beeplant**. Because of the unpleasant odor of crushed herbage, this plant is sometimes called **Skunkweed**. The flowers are an important source of honey, and the seeds are eaten by a number of birds, especially doves.

GREENLEAF MANZANITA

Arctostaphylos patula   ERICACEAE | HEATH FAMILY

Many people are attracted to **Greenleaf Manzanita** by its bright mahogany-red bark. Its oval-shaped leaves are a bright green throughout the year. The flowers grow in clusters and sometimes are very numerous on the shrub. The fruit resembles a tiny apple, and the name manzanita is Spanish for "little apple." Native Americans use the berries for food and for making a pleasant, sour drink.

RATTLEWEED, MILKVETCH, POISON-VETCH

Astragalus sabulonum   FABACEAE | PEA FAMILY

This showy species of *Astragalus* is locally called **Rattleweed** because when it is in fruit, its large, bladder-like, thin-walled pods become very brittle and give a distinct rattling sound when shaken. The pods are about 1.5 inches long and heavily mottled reddish-brown in color. All *Astragalus* have pinnately compound leaves, with the leaflets usually rather closely crowded together. Many *Astragalus* are poisonous to grazing animals, but some seem to be comparatively harmless. The species poisonous to livestock are commonly called **Loco Weeds** or **Poison-Vetches** (which prefer soils rich in selenium, and take up enough of that toxic mineral to make them poisonous to stock, especially sheep). The harmless species are often called **Milkvetches**. Nearly all the species are colorful and spectacular when in blossom, but some of them have a rank, disagreeable odor.

This very large genus of plants ranges from the hottest parts of the desert to high mountain peaks and far to the North. More than a dozen species are found in the Zion region, especially at lower elevations.

NEW MEXICO LOCUST

Robinia neomexicana   FABACEAE | PEA FAMILY

This shrub or small tree is fairly common in Zion Canyon, and some plants were probably brought in by early settlers. The large, showy flowers, blooming in May and June, grow in clusters at the ends of slender branches. The tree is very thorny and has the habit of sprouting from roots or stumps and of forming dense thickets which are valuable in controlling erosion. The foliage serves as food for browsing animals, especially deer.

ALFILERIA

Erodium cicutarium　　　　　GERANIACEAE | GERANIUM FAMILY

This low-growing, introduced plant, spreading close to the ground, with its finely divided leaves and small, starry-pink flowers, puts on a remarkable display in the open meadows of the large canyons. It is one of the earliest blooming species in Zion Canyon, and in seasons of abundant rain, it often presents the appearance of a pale-purple lawn. On ripening, the seed capsules split open and shoot out the seeds, each seed with a tiny hook in its nose and a tail with successive tight coils like a corkscrew. The seed is apparently screwed into the ground by alternating moisture and dryness which winds and unwinds the seed plume.

FREMONT GERANIUM

Geranium caespitosum   GERANIACEAE | GERANIUM FAMILY

This beautiful, midsummer-blooming plant, growing about two feet high, is common on the plateaus and in the cool canyons. The pink, veined petals, deeply lobed leaves and characteristic geranium odor help identify this plant. Some species have white flowers, but they are not common in this area. The flowers are perfect, with five sepals, five petals, and five to ten stamens. The fruit is a long capsule and has given rise to the common name **Cranesbill**. Note that our cultivated Geraniums are really members of the genus *Pelargonium* from South Africa.

COLORADO FOUR-O'CLOCK

Mirabilis multiflora   NYCTAGINACEAE | FOUR-O'CLOCK FAMILY

Closely resembling the cultivated variety of Four-O'Clock, this plant, with its abundance of brilliant magenta-colored flowers, is one of the spectacular wildflowers of May or early June. It is a sturdy perennial with thick, glossy-green leaves spreading low over the ground. The south-facing slopes in the Sonoran Zones are its most common habitat, but it is also found in the broken lava fields. Being a night-bloomer, the flowers close during the bright daylight hours and then open at about four o'clock in the afternoon. Its blooming season is generally brief, about two or three weeks, but it sometimes blooms twice in the same summer.

FIREWEED

Chamaenerion angustifolium ONAGRACEAE | EVENING-PRIMROSE FAMILY

SYNONYM *Epilobium angustifolium*

This tall, willowy herb is frequently the first plant to come in after a forest fire. Its colorful bloom gives new life to the blackened ground. **Fireweed** is one of the world's most widely disseminated wildflowers, being found throughout much of northern North America, Europe, and parts of Asia. Its seeds are scattered by the wind. In our area it is found mostly on the high plateaus.

PALMER PENSTEMON

Penstemon palmeri   PLANTAGINACEAE | PLANTAIN FAMILY

This is one of the very beautiful and conspicuous flowers of Zion National Park. The flowers are borne on tall spikes and are brightly colored. **Palmer Penstemon** is the largest and most common Penstemon found along the trails and roadways of Zion below 6,000 feet. It is especially common in freshly disturbed soils on road-cuts. The leaves are gray-green, and each pair (with the exception of the lowermost) is joined at the base, creating the impression that it is one leaf with the stem growing through the center. The large flowers, which are pale lavender in color, begin blooming in May, and are found in bloom throughout most of the summer.

PARRY PRIMROSE

Primula parryi

PRIMULACEAE | PRIMROSE FAMILY

This attractive plant is found only at high elevations, generally above 10,000 feet. In this region it grows fairly abundantly on Brian Head Peak but is found sparingly at Cedar Breaks. Its brilliant display of rose-red flowers is a remarkable and rewarding sight for those who gain the high places and see this alpine beauty. The smooth, thick leaves, which are quite long, all grow in a rosette at the base of the plant. The fragrance of this flower is disappointing, however, for it does not match its splendid color.

SHOOTINGSTAR

Primula pauciflora   PRIMULACEAE | PRIMROSE FAMILY

SYNONYM *Dodecatheon pauciflorum*

Shootingstar is one of the early blooming flowers in the alpine meadows of Cedar Breaks and on the high plateaus. They also develop very early in the moist canyons of Zion. Along with **Columbine** and **Monkeyflower** they are the predominant plants of the 'Hanging Gardens' found on many of the canyon walls. The basal leaves spread close to the ground, while the flowers, in a variety of colors – white, pink or purple – grow on stems 6 to 8 inches high. The down-pointed stamens of the flower center and the reflexed or turned-back petals gives the flower its common name of **Shootingstar**.

SAND BUTTERCUP

Ranunculus andersonii   RANUNCULACEAE | BUTTERCUP FAMILY

SYNONYMS *Beckwithia andersonii, Ranunculus juniperinus*

Sand Buttercup is one of the very early blooming plants in Zion in the Transition or Upper Sonoran Zone, often appearing from mid-February to April. It grows in bare sandy or rocky places among the Junipers and Pinyons. Along the trail to the Canyon Overlook above the Great Arch is perhaps the best place to find this plant in Zion. The flowers, a pinkish white, are found on short stems or spreading branches held closely to the ground. As the plants are rather small and not very showy, it takes careful searching to find them.

PRAIRIESMOKE

Geum triflorum ROSACEAE | ROSE FAMILY

This graceful plant, with its nodding, bell-shaped, pink-colored flowers, is found fairly abundantly in the alpine meadows of Cedar Breaks National Monument. The plant has a number of common names such as: **China Bells**, **Oldman-Whiskers**, and **Grandfather's Beard**. The silvery, plumose tails of the fruit present an attractive display, especially as the sun's rays light the waving plumes in late afternoon or early morning. Prairiesmoke plants are considered good forage for several animals, and bumblebees gather its pollen for honey.

WILD ROSE

Rosa spp.   ROSACEAE | ROSE FAMILY

There are several species of **Wild Rose** in Zion, Bryce Canyon, and Cedar Breaks National Monument; illustrated above is **Woods' Rose**, *Rosa woodsii*. Wild Roses are widely distributed in the northern hemisphere and are too familiar to need much description. The flowers are fairly fragrant and have bright pink petals with a large cluster of yellow stamens. The fruit or 'hip' of the rose, rich in Vitamin C, and shaped like a small apple, turns a deep-red color late in the season and adds beauty to this plant during autumn.

BUTTERFLY MILKWEED

Asclepias tuberosa   APOCYNACEAE | DOGBANE FAMILY

There are a handful of fairly common species of Milkweed in Zion, but **Butterfly Milkweed** is the most common. This plant is found in dry places above 4,000 feet and is especially abundant in Birch Creek Canyon. The conspicuous orange flowers grow on fairly tall stalks about two feet in height and make this plant very easy to find. The stems are quite hairy, leafy and contain a milky juice. As the fruits develop in large boat-shaped pods, the seeds burst forth bearing long, silky hairs that assist the wind in scattering them over wide areas.

WESTERN CARDINALFLOWER

Lobelia cardinalis　　　CAMPANULACEAE | BELLFLOWER FAMILY

A pleasant surprise to many park visitors is to find this spectacular flower, with its abundant scarlet blooms on long stalks, presenting a colorful display during the late summer when most other plants have ceased blooming. This colorful species, known also as **Scarlet Lobelia**, is very abundant along the Narrows Trail of Zion Canyon, and also along watercourses of other shady canyons. The long, tubular corollas and pointed petals arranged in an irregular pattern of two and three separate this plant from the **Scarlet Penstemon** which it resembles.

SCARLET GLOBEMALLOW

Sphaeralcea coccinea   MALVACEAE | MALLOW FAMILY

Scarlet Globemallow is very common along roadsides, and especially prominent in campgrounds or other disturbed areas. Plants have a lovely coral-red display as early as May, and they continue blooming throughout much of the summer. Leaves are palm-like, divided into 3 or 5 leaflets, each split into two or more rounded or pointed segments; and the leaves and stems are covered with soft, velvety hairs. **Cotton** (*Gossypium*) belongs to this large and important family, which also contains such ornamental plants as the **Hollyhock**. Extensive fields of Globemallow provide a brilliant display of orange.

HUMMINGBIRD TRUMPET

Epilobium canum   ONAGRACEAE | EVENING-PRIMROSE FAMILY

SYNONYM *Zauschneria garrettii*

One of the late blooming plants in Zion National Park is **Hummingbird Trumpet**, also called **Fire-Chalice**, or **Wild Fushia**. It can be found on the Canyon Overlook Trail or on the West Rim Trail at elevations near 6,000 feet. It is identified by the narrow oval leaves pointed and toothed, and the fushia-like flowers, narrowly funnel-shaped, with the pistil and stamens extending beyond the petals. The brilliant scarlet of this flower grouped in fairly dense clusters makes an attractive display in late August and September.

DESERT INDIAN-PAINTBRUSH

Castilleja chromosa OROBANCHACEAE | BROOM-RAPE FAMILY

From early March until May, warm hillsides below 6,000 feet are brilliantly colored by clumps of deep-red flowers (often found next to patches of **Mountain Mahogany**). They are the **Desert Indian-Paintbrush**, a very conspicuous early spring flower in Zion. Found abundantly along the park road from the East Entrance to the Zion Tunnel, they present a most pleasing sight early in the season. Other species of Indian-Paintbrush are plentiful at Cedar Breaks, and often carpet the meadows in showy orange or red. Several species are also found at Bryce Canyon National Park.

MONKEYFLOWER

Mimulus eastwoodiae   PHRYMACEAE | LOPSEED FAMILY

SYNONYM *Mimulus cardinalis*

One of the very beautiful flowering plants along the Zion Narrows Trail and in cool, damp places of the shady canyons is this crimson **Monkeyflower** with its orange-red blossoms and deep-green leaves. Its flowers are 1 to 2 inches long, and the wide-toothed leaves are 3 to 5 inches long. It is the largest Monkeyflower in the Park. Some plants are found blooming throughout much of the summer season, especially along the canyon walls where there are seeps of water most of the year.

EATON PENSTEMON

Penstemon eatoni   PLANTAGINACEAE | PLANTAIN FAMILY

Found mostly in the cool canyons, this plant is sometimes confused with **Skyrocket Gilia** or **Western Cardinalflower**, which it resembles somewhat. Penstemon usually has a greater number of blooms on each flower stalk than do these flowers that appear like it. **Eaton Penstemon** is not nearly as common as many other Penstemons in these areas. However, it is far more brilliantly colored which accounts for such common names as **Firecracker Penstemon** and **Scarlet Penstemon**.

SKYROCKET GILIA

Ipomopsis aggregata　　　　　POLEMONIACEAE | PHLOX FAMILY

SYNONYM *Gilia aggregata*

This plant is found most commonly in the Ponderosa Pine belt where its star-shaped, scarlet flower adds a bit of brilliance to the landscape. The individual flowers, with their long, tubular corollas and star-shaped petals, are master-pieces of beauty. Their shape and color have given rise to other common names as **Trumpet Phlox** and **Scarlet Gilia**. In their search for nectar, hummingbirds are noticeably attracted to the flowers of the plant. Birds and insects, by taking the nectar, help in the pollination of many flowers.

HEARTLEAF ARNICA
Arnica cordifolia ASTERACEAE | SUNFLOWER FAMILY

A common flower in the Pine and Spruce forest of Cedar Breaks, and in alpine areas of Zion and Bryce Canyon, is **Heartleaf Arnica**. The yellow ray flowers are few, while the disk or central flowers of the flower head are numerous. The flowers measure about three inches across and are often mistaken for sunflowers. The heart-shaped leaves help distinguish this flower from its close relatives. Tincture of arnica is obtained from certain species of Arnica.

DESERT MARIGOLD

Baileya multiradiata　　　　ASTERACEAE | SUNFLOWER FAMILY

These golden-yellow flowers, measuring about three inches across, are fairly common along the trails and roadways of Zion Canyon and in other low-elevation areas of the park. They bloom during May and June. The ray flowers become bleached and papery as they mature, thus accounting for the name **Paper Daisy**. This attractive composite is also known as **Desert Baileya**. In California this plant is cultivated for the flower trade. It has been found poisonous to sheep, although horses crop the flower heads, apparently without harmful effects.

ARROWLEAF BALSAMROOT

Balsamorhiza sagittata ASTERACEAE | SUNFLOWER FAMILY

Arrowleaf Balsamroot, with its large, showy yellow flowers, is often found on southern exposures of steep hillsides or in sagebrush flats. It was first described by Lewis and Clark on their expedition across the continent in 1804–1806. The rind of the root contains a turpentiny balsam, but the heart of the root is edible and was used by Native Americans and early pioneers. The plant is called **Mormon Biscuit** in Utah. The seeds of the plant were used by Native Americans to make "Pinole" or meal, and the stems and leaves were eaten as greens.

RUBBER-RABBITBRUSH

Ericameria nauseosa　　　　　ASTERACEAE | SUNFLOWER FAMILY

SYNONYM *Chrysothamnus* spp.

Illustrated here is one of the region's ten species of rabbitbrush, *Ericameria nauseosa*, especially common in the Sonoran Zones; the related *Ericameria parryi* occurs in the Transition Zone. Because rabbits find this plant a favorite shelter, it has been named **Rabbitbrush**. Native Americans boil the plant for yellow dye, and early settlers found certain species of *Ericameria* to contain rubber. Consideration was given to the production of rubber from this species during the First World War and up until the discovery of synthetic rubber.

GAILLARDIA, BLANKET-FLOWER
Gaillardia pinnatifida   ASTERACEAE | SUNFLOWER FAMILY

This handsome and conspicuous plant is found growing mostly in the Sonoran Zone. It has a slender, rough stalk, about a foot tall, and stiff, rather hairy, dull-green, incised leaves growing mostly from the root. The beautiful flowers, about three inches across, have golden-yellow rays which are three-tipped. The center of the flower is a velvety maroon and yellow, becoming an attractive fuzzy, round, purplish head when the rays are shed. **Gaillardia** blooms mostly in May and June.

CURLYCUP GUMWEED

Grindelia squarrosa   ASTERACEAE | SUNFLOWER FAMILY

Curlycup Gumweed is a native but weedy plant, probably spread by vehicles, as it is found most commonly along roadways or in culitivated fields. Once established, it spreads rapidly along highways or in cultivated areas. The plant is suspected to be toxic to livestock, but is rarely eaten as it has a sticky coating. It was formerly used in the treatment of asthma in humans. In addition, external use is made of it to relieve the irritation caused by **Poison Ivy**.

MOUNTAIN SUNFLOWER

Helianthella uniflora　　　　ASTERACEAE | SUNFLOWER FAMILY

Sunflowers abound in the Parks during the late summer, and a common species, **Mountain Sunflower**, is illustrated here. As early summer flowers (mostly in blues and purples) fade, yellow and red flowers become prominent. This is especially true in the alpine meadows of Cedar Breaks in August when one of the dominant plant groups are the **Sunflowers**. The seeds of Sunflowers supply abundant food for many birds and small mammals.

COMMON SUNFLOWER

Helianthus annuus ASTERACEAE | SUNFLOWER FAMILY

The very large flowers of these plants sometimes present a colorful display as they take over roadsides or abandoned fields. Members of this group are often considered weeds because of their habit of crowding out more desirable species. Certain species of **Sunflower** have been developed for commercial purposes, and produce oil for cooking and meal for livestock feed. Native Americans in some areas of North America cultivated sunflowers for food and for trade.

WESTERN CONEFLOWER

Rudbeckia occidentalis　　　ASTERACEAE | SUNFLOWER FAMILY

At Cedar Breaks and in the high elevations of Zion and Bryce Canyon you can find this rank-growing plant in fair abundance. Its thimble-like, dark-brown flower head has numerous, inconspicuously small yellow flowers that bloom progressively up the cone from its base. The dark-brown cones, held above the foliage of the plant, make a spectacular display against the deep-blue sky. The ripened seeds are sought after by rodents and numerous birds.

FREMONT BARBERRY

Berberis fremontii BERBERIDACEAE | BARBERRY FAMILY

SYNONYM *Alloberberis fremontii, Mahonia fremontii*

This rather tall shrub of the Sonoran Zones puts on a remarkable display in April and May with its bright yellow flowers. It is common along the highway leading to Zion National Park from the west. Since the plant is a secondary host of the blackstem rust of cereals, it is not cultivated as an ornamental shrub. Native Americans used the wood of **Fremont Barberry** for various implements or tools. They utilized the root, which contains berberine, for a tonic, and they also made from it a brilliant-yellow dye.

OREGON GRAPE

Berberis repens　　　　　BERBERIDACEAE | BARBERRY FAMILY

SYNONYM *Mahonia repens*

Because of its hollylike leaves, this dwarf shrub is sometimes called **Holly Grape**, but it is more commonly known as **Oregon Grape**. A rather prostrate growth form accounts for a third common name of **Creeping Barberry**. This plant is found sparingly scattered in the region, and is probably more abundant in Zion than in Bryce Canyon or Cedar Breaks. The fruit looks very much like a cluster of grapes and is gathered for the making of jellies or wine. The woody stems were used by Native Americans in making a yellow dye. The plants are helpful in holding the soil, as they spread close to the ground.

NARROWLEAF PUCCOON

Lithospermum incisum   BORAGINACEAE | BORAGE FAMILY

Narrowleaf Puccoon is well adapted to dry habitats. The plants are commonly found in clumps, but they are generally widely scattered rather than in dense growths as in the case of the **Bluebells** or **Mertensias** that belong to the same plant family. Its showy, trumpet-like yellow flowers attract many insects as they bloom during April and May. The seeds are hard, white and shiny, hence the name *Lithospermum*, meaning 'stone seed.'

DESERT PRINCE'S-PLUME

Stanleya pinnata   BRASSICACEAE | MUSTARD FAMILY

During the months of May and June the very conspicuous **Desert Prince's-Plume** in Zion Canyon and throughout the Sonoran Zones may be found sending up its tall spikes of lemon-yellow flowers. On the same stalk and at the same time can be found the ripened and opened seed pods (siliques), fresh-blooming flowers and unopened buds. It has tall, stout stems, rather woody at the base, and differs from many plants in that it is tolerant of soils containing high concentrations of gypsum (calcium sulfate hydrate).

WESTERN WALLFLOWER

Erysimum capitatum   BRASSICACEAE | MUSTARD FAMILY

There are several kinds of both native and introduced **Wallflowers** in Zion National Park. Their bright yellow flowers, which grow on stalks taller than those of most other mustards, make them among the most attractive members of this family. They are usually found on rather dry slopes in the Upper Sonoran and Transition Zones. Notice how the petals are arranged as a cross which is a characteristic of all members of the Mustard Family.

BEARBERRY HONEYSUCKLE

Lonicera involucrata   CAPRIFOLIACEAE | HONEYSUCKLE FAMILY

This member of the Honeysuckle Family has a number of common names such as **Twinberry Honeysuckle, Ink-Berry** and **Pigeon-Bush**. The flowers, which are yellow and always come in pairs, are very attractive to hummingbirds. Mature fruits are black berries about the size of peas and are partially enclosed by reddish bracts. The plants are unpalatable and browsed very slightly, but the fruits are eaten by birds and chipmunks.

YELLOW SPIDERFLOWER

Peritoma lutea CLEOMACEAE | SPIDER-FLOWER FAMILY

SYNONYM *Cleome lutea*

The plants of this genus are often called **Beeplants**. There are two species of Spiderflower in the region. **Yellow Spiderflower** is not quite as common as the purplish-pink species commonly known as **Rocky Mountain Beeplant** (*Peritoma serrulata*). Both species are conspicuous roadside flowers in June and July. Although they are important sources of honey, they are not very sweet-scented to humans. No doubt their odor helps attract insects to the flowers.

STONECROP

Sedum lanceolatum   CRASSULACEAE | STONECROP FAMILY

SYNONYM *Sedum stenopetalum*

Found mostly in very dry, rocky soil, these small plants, with smooth, fleshy leaves and star-like yellow flowers, are fairly conspicuous as they bloom during the early summer months. These plants have the ability to store moisture in their fleshy leaves and stems. They are, therefore, well adapted to withstand long periods of drought. **Stonecrop** is sometimes gathered for treatment of certain ailments.

ROUND-LEAF BUFFALOBERRY

Shepherdia rotundifolia ELAEAGNACEAE | OLEASTER FAMILY

Round-Leaf Buffaloberry is a low, evergreen shrub. Its leaves are small, oval, and appear to have been covered with a thin coat of aluminum paint through which the green shows faintly. It is most common in the Upper Sonoran Zone and may be found along the Canyon Overlook and Emerald Pool Trails in Zion National Park. The pale-yellow flowers, to about ¼ inch across, are often hidden by the leaves. They bloom in early April (or sometimes in March). The fruit of a similar shrub has a tart berry that was gathered by early pioneers and used as a sauce on Buffalo steaks, hence the name Buffaloberry.

HONEY MESQUITE

Prosopis glandulosa

FABACEAE | PEA FAMILY

SYNONYM *Prosopis juliflora*

Honey Mesquite is a low-growing tree of the Lower Sonoran Zone, uncommon in Zion but fairly abundant in the desert area adjacent to the Park. In early spring (March and April), bright-green leaves cover the tree, and often it is laden with catkin-like clusters of greenish-yellow flowers, which attract a host of insects including honey bees. Early settlers made extensive use of the wood of this tree for fuel, building corrals, and in making furniture and utensils. The fruit of the **Honey Mesquite**, resembling a string bean, is eaten by many animals. Native Americans also made wide use of it by grinding the beans into a meal called "Pinole."

YELLOW MARIPOSA

Calochortus aureus

LILIACEAE | LILY FAMILY

Found only in the petrified forest of the Coalpits Wash section of Zion National Park, this plant is not very abundant and probably declined from overgrazing by livestock during the settlement period before Zion became a national park. This species, with bright-yellow flowers, is associated with a specific geologic stratum – the Petrified Forest member of the Chinle formation. **Yellow Mariposa** is found in great abundance in the Petrified Forest National Park near Holbrook, Arizona. Mariposa in Spanish means 'butterfly.'

PURPLESPOT FRITILLARY

Fritillaria atropurpurea   LILIACEAE | LILY FAMILY

A rather rare lily sometimes also called **Leopard Lily** or **Bronze Bell**. Plants with their drooping flowers on fairly tall stems are found in Sagebrush areas or in alpine meadows. As they are not very conspicuous, they are often over-looked by visitors. The petals, with their mottled effect of brown, yellow and purple spots, present a remarkable pattern of beauty when observed closely. The odor of the plant is not pleasing to people, but is no doubt attractive to insects.

DESERT BLAZINGSTAR

Mentzelia integra

LOASACEAE | BLAZINGSTAR FAMILY

SYNONYM *Mentzelia multiflora*

Blazingstar, sometimes called "Stickleaf" because of its rough, hairy leaves, is well-suited to the drought conditions of this region, as they seem to prefer dry, rocky soil. They are often found in roadside cuts or other newly disturbed soils. Plants typically bloom in July and August. They are conspicuous with their yellow flowers consisting of five long petals and a large number of stamens almost as long as the petals that attract the eye in the bright sunlight of midsummer. **Desert Blazingstar** is found mostly in the Transition Zone.

YELLOW EVENING-PRIMROSE

Oenothera villosa ONAGRACEAE | EVENING-PRIMROSE FAMILY

SYNONYM *Oenothera strigosa*

Yellow Evening-Primrose is commonly found in road cuts or in places where the soil has been disturbed. This species and other members of this family have very showy flowers with four broad, thin petals. Generally the plants bloom at night but sometimes in the daytime if growing in deep shade. The **Evening-Primroses** are among the comparatively few flowers blooming in Zion Canyon during the heat of midsummer, and many of them are usually found on sandy or rocky soil in the Upper Sonoran zone.

WILD BUCKWHEAT

Eriogonum umbellatum   POLYGONACEAE | BUCKWHEAT FAMILY

Wild Buckwheat is commonly associated with Sagebrush and with arid regions of the West. Many species of this large genus are found blooming throughout the summer. The spreading branches grow close to the ground and help reduce erosion, and the yearly accumulation of leaves adds humus to the soil. The flower head at the top of single stalks, with its many-branched, dense cluster in a lacy pattern, makes a fine floral display of yellow. The flowers are important to honey bees; and ripened seeds are sought by chipmunks, other rodents, and several types of birds.

MARSH-MARIGOLD

Caltha leptosepala　　RANUNCULACEAE | BUTTERCUP FAMILY

Marsh-Marigold is a plant of high elevations, generally above 9,000 feet, and found abundantly at Cedar Breaks, where it comes into bloom almost as soon as the first patches of bare ground appear in spring (usually in April or May). **Marsh-marigolds** often carpet the alpine meadows with a spread of white blossoms. The white sepals, that make up the showy flower, are often mistaken for petals, which are absent. The mass of anthers of the stamens give the flower its brilliant yellow center.

BUTTERCUP

Ranunculus spp.

RANUNCULACEAE | BUTTERCUP FAMILY

A number of species of **Buttercup** bloom in the region in early April or May on the Plateaus, and later in the shady canyons. Shown above is **Sagebrush Buttercup** (*Ranunculus glaberrimus*). They are often the first flowers of spring, followed closely by **Springbeauty** or sometimes preceded by the tiny white-flowered **Indian-Potato** of the Carrot Family. A notable sight is to find the waxy flowers of the early Buttercups at the very edge of the receding snowbanks. The blooming season for Buttercups is very brief, as a general rule, but the different species come into bloom successively.

LITTLELEAF MOUNTAIN-MAHOGANY

Cercocarpus ledifolius   ROSACEAE | ROSE FAMILY

This low-growing shrub is fairly important as winter browse for deer and other browsing animals. The leathery leaves are evergreen, rather narrow, pointed at both ends and curled downwards along their sides. The flowers are very small and inconspicuous, but the fruits, with their long plumes, present an interesting display. The dead wood of this shrub is very useful to campers, as it burns with an extremely hot flame and gives off very little smoke.

BLACKBRUSH

Coleogyne ramosissima　　　　ROSACEAE | ROSE FAMILY

Blackbrush is a shrub found mostly in the Sonoran Zones of Zion National Park. Several plants may be seen near the South Entrance Station. It is well named, as it has a burned and dead appearance during much of the year; however in late April and May it puts out minute gray-green leaves and creamy-yellow flowers made up of four sepals and no petals; the stamens are numerous. **Cliffrose, Bitterbrush** and **Mountain-mahogany** are closely related to the Blackbrush.

BUSH CINQUEFOIL

Dasiphora fruticosa

ROSACEAE | ROSE FAMILY

SYNONYM *Potentilla fruticosa*

Found most commonly at Cedar Breaks, **Bush Cinquefoil** puts on a very showy display for a brief period of the summer, generally in July (or earlier at lower elevations). It is a shrub commonly 2 to 3 feet high, with leaves that are five-parted, hence the name **Cinquefoil**. The shrub is often dotted all over with rose-like flowers, about an inch across, with yellow petals and deeper yellow anthers. The plant is common in the mountains, across the continent, and up to altitudes of 10,000 feet or more.

ANTELOPE BITTERBRUSH

Purshia tridentata   ROSACEAE | ROSE FAMILY

This shrub looks much like **Cliffrose**, except for size. It is lower growing, has smaller pale-yellow flowers, and its fruits are not plumed. The stubby, fan-shaped leaves are three pointed like those of **Big Sagebrush**. Bitterbrush is found most commonly on warm, dry slopes above 4,000 feet, where it provides valuable year-round browse for deer and other animals. It is also known as **Antelopebrush** and **Quininebush**.

MULLEIN

Verbascum thapsus   SCROPHULARIACEAE | FIGWORT FAMILY

This rather unusual-appearing plant is not a native but rather an introduced species from Europe that has spread over most of North America. In the Zion region, it is fairly common along roads and trails. The Hopi are said to dry and smoke the leaves, and this was thought to cure people who are mentally unbalanced. Early Greeks and Romans dipped dried mullein stalks in tallow to make lampwicks. The English named it "Torchweed" and the Spanish called it "Candelaria."

CREOSOTEBUSH

Larrea tridentata   ZYGOPHYLLACEAE | CREOSOTE-BUSH FAMILY

Probably the most characteristic shrub of the arid valleys and mesas of the Southwest is **Creosotebush**, which is sometimes erroneously called 'Grease-wood' (*Sarcobatus*). At times, this plant fills the air with a peculiar pungent aroma that gives rise to the common name **Creosotebush**. Mexicans call it "Hediondilla" (the 'Little Bad Smeller'). Traditionally, a sticky resin on the leaves was used as a poultice for bruises and sores. Also, a resinous gum or 'lac' deposited by scale insects on the branches was used by Native Americans as a cement for fixing arrow points and for mending pottery.

BUCKHORN CHOLLA CACTUS

Cylindropuntia acanthocarpa   CACTACEAE | CACTUS FAMILY

SYNONYM *Opuntia acanthocarpa*

The Cactus Family is well represented in the region, where at least ten different genera are represented; six of the more common species are illustrated in this guide. **Cholla** (pronounced 'Choya') **Cactus** is the only "tree" cactus found in Zion. It is fairly common in the Lower Sonoran Zone, especially in Coalpits Wash. Strangely enough, this plant is a favorite nesting place of some desert birds, including sparrows, wrens, and finches. However, the spreading branches seem to reach out and grab the careless hiker, and the barbed spines stick so fast in the flesh that the joint of the plant is separated from the main stem before the spines can be withdrawn. **Cholla** blossoms come in many shades of color, most commonly in yellows and reds.

SCARLET HEDGEHOG CACTUS

Echinocereus coccineus   CACTACEAE | CACTUS FAMILY

This species is sometimes called the **Cucumber Cactus**. It is found blooming in bright-red clumps as early as April or on occasion in March when the winters are not severe. Its favorite habitat appears to be the rocky slopes of lava fields below 5,000 feet. As cactus flowers mature into fruit, they form bulb-like bodies called *tunas*. The fruits of this species are about an inch or more in diameter and serve as important food for many rodents.

PURPLETORCH CACTUS

Echinocereus engelmannii CACTACEAE | CACTUS FAMILY

Found mostly on rocky slopes in the Lower Sonoran Zone, **Purpletorch Cactus** is fairly common, and distinguished chiefly by its waxy, brilliantly colored purple flowers. They generally bloom during the month of May.

BEAVER-TAIL CACTUS

Opuntia basilaris   CACTACEAE | CACTUS FAMILY

One species of cactus that at first sight does not seem to be a cactus is the Beaver-Tail, for it lacks the long spines characteristic of most cacti. Upon close examination, however, you will find that it is protected by numerous short, fine spines. The name beaver-tail is derived from its flat stems shaped somewhat like a small beaver's tail. The conspicuous magenta flowers mark this species as one of the most beautiful in the Park. The fruit of this species is an important animal food, especially for chipmunks and ground squirrels.

ENGLEMANN PRICKLY-PEAR CACTUS

Opuntia engelmannii   CACTACEAE | CACTUS FAMILY

As one of the largest of the flat-jointed or prickly-pear cacti in Zion, this species is fairly abundant in Zion Canyon, and is probably the most frequently observed species because of its dense growth in certain habitats, its sometimes immense size, and its colorful display of blossoms. It is locally called **Elephant Ear Cactus**. In June it produces large, yellow, waxy-textured flowers.

Engelmann Prickly-Pear Cactus, left, Beaver-Tail Cactus, right.

Red fruit of **Engelmann Prickly-Pear Cactus**, locally called 'Tuna' or 'Cactus Apple'.

The fruit of the **Engelmann Prickly-Pear Cactus** is similar to the fruits of many other species of cacti; it is conspicuous, being deep red in color and about the size of a large crab apple. It is called '**Tuna**' or locally '**Cactus Apple**,' and ripens as early as June but more abundantly during July. Native Americans of the Southwest use this fruit as an important item of their diet. Many people living in the desert have come to learn that this cactus fruit makes excellent jellies and candies. It is also feasted upon by several rodents, especially the Antelope Ground Squirrel.

PRICKLY-PEAR CACTUS

Opuntia polyacantha

CACTACEAE | CACTUS FAMILY

SYNONYM *Opuntia rhodantha*

The most common cactus in Zion is this species of **Prickly-Pear Cactus**. It is found in a great variety of habitats, even at higher elevations. The flowers, large and spectacular in various shades of red, salmon, or yellow, bloom in late May to July. Because of its long blooming season, this species is more often found in flower than any other cactus in the Park. Its pear-shaped fruit, red to purple in color, is eaten by many animals as well as by the native peoples.

INDEX (PLANT FAMILY)

INDEX

Common names are shown in **bold**; accepted scientific names in roman, synonyms are listed in *italics*.

CPSIA information can be obtained
at www.ICGtesting.com
Printed in the USA
LVHW070340110222
710561LV00010B/648